Challenging Math Problems

TERRY STICKELS

D1157476

DOVER PUBLICATIONS, INC.
Mineola, New York

Copyright

Bibliographical Note

Challenging Math Problems is a new work, first published by
Dover Publications Inc., in 2015.

International Standard Book Number
ISBN-13: 978-0-486-79553-9
ISBN-10: 0-486-79553-5

Manufactured in the United States by RR Donnelley
79553501   2015
www.doverpublications.com

CONTENTS

INTRODUCTION

It is the wish of every writer to write that one book he/she has always wanted to see on book shelves. From what I have seen and read, it doesn't happen often. The book you now hold in your hands is that book for me. For over eight years, I put together the best and most challenging puzzles I could, hoping a good publisher would like it as much as I did. I kept them in a file and reviewed them weekly, often asking for advice and suggestions from some very bright people. When I had a discussion with Mr. John Grafton of Dover Publications about a possible book of puzzles, I was pleasantly surprised when he asked if I would be interested in doing a book of the most challenging puzzles imaginable. The puzzles had to be the kind that require no higher math ... just good thinking skills. And they had to be fun. They had to offer that AHA! moment of discovery that all puzzle solvers love. That was good for me because it is difficult to take myself or puzzles too seriously.

Regarding the aspect of having fun, I owe my outlook to the best puzzle writer, Martin Gardner. He got a kick out of everything he wrote about and puzzles were at the top of his list. During our discussions he would always mention some topic or puzzle that would make him laugh, followed by saying, "I have never had to work a real job a day in my life. I've been having fun my whole adult life." How many people can say that?!

So, if I have done my job properly, you will have a good time solving some demanding puzzles. That's it. That is what I set out to do: keep things neat, straight-ahead, and fun. That is my reward for writing puzzles for a living.

ACKNOWLEDGMENTS

The following people have bent over backward for years helping me, often working weekends and nights just to make a puzzle perfect. Diana Smith is the editor for my *Stickelers* puzzle column syndicated by King Features. Her suggestions and keen eye catching my errors are invaluable. Terry Baughan is my graphic artist/designer for *Stickelers* and she takes painstaking efforts to make sure each graphic is the way I want it, a very difficult job. Christy Davis has been my associate for many years and is responsible for putting everything together. She is a wonderful puzzle writer, as well. Her Word Search puzzle books and calendars are great fun.

Special mention to two men who are both brilliant mathematicians and friends: Dr. John Konvalina, Department Chair of Mathematics at the University of Nebraska at Omaha and Dr. Harvy Baker of the Mathematics Department at the University of

Texas – Arlington. Both spent countless hours helping refine the puzzles in this book, making suggestions and explaining concepts in terms that I could, in turn, explain to the reader. They did this for the love of learning and for the fun of solving puzzles.

Thank you all.
Terry Stickels
www.terrystickels.com

PUZZLES

1

If a teacher can place his students eight to a bench, he will have only three students on the final bench. If he decides to place nine students on a bench, he'll have only four students on the final bench. What is the smallest number of students this class could have?

2

Here's a probability puzzle that may test even some of the best mathematicians.

Imagine you have two opaque boxes. One box has one white marble and the other box has one white marble and one black marble. Of course, you can't see into either box. Simultaneously, you reach into both boxes, grab one marble from each, and quickly switch the marbles without looking at them. You then pick one of the marbles from one of the boxes. What is the probability the marble you pick is white?

3

Molly places 200 kilograms of watermelons in cases in front of her shop. At that moment, the watermelons are 99% water. In the afternoon, it turns out that it is the hottest day of the year, and as a result, the watermelons start to dry out. At the end of the day, Molly is surprised because she wasn't able to sell a single watermelon, and the melons are now only 98% water. How many kilograms of watermelons does Molly have left at the end of the day?

4

One of the most interesting studies in mathematics is the way numbers grow as they approach infinity. So, for this puzzle consider if you were to take the limit, as "n" approaches infinity, put the following in order, going from lowest to highest in value. Note: "c" is any positive integer you choose greater than 1. All values of "n" are positive and greater than one.

$$n^c \qquad \ln n \qquad n! \qquad n \qquad c^n \qquad n^n$$

5

What number comes **before** the 3 to start this sequence?

? 3 11 31 75 155 283 471

6

A positive whole number is divisible by 3 and also by 5. When the number is divided by 7, the remainder is 5. What is the smallest number that could work?

7

What are the missing digits in the division below? All the 3s are given. The placeholder Xs may be any digit from 0 – 9, excluding 3.

```
              3 X X X
        33) X X X X X 3
            X X
            3 X X
            X X X
                3
```

8

In the figure below, the total area of rectangle BCDE plus triangle ABE is 192 sq. units. What is the length of AB?

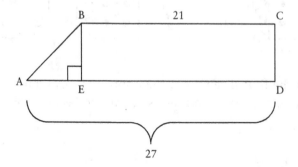

9

The mean of three numbers is 12 more than the least of the numbers and 18 less than the greatest. The median of the three numbers is 9. What is the sum of the three numbers?

10

Alison and Amelia live 14 miles apart. Alison started to drive her car toward Amelia's house. At the same time, Amelia started to drive toward Alison's house. When they met, Alison had been driving for three times as long as Amelia and at 3/5 Amelia's rate. How many miles had Amelia driven when they met?

11

On a normal die, the sum of the dots on opposite faces is seven.

A   B   C   D

On the four dice above, the opposite faces total seven but one of the pictures is incorrect because of the orientation of the dots. Which die is incorrect? There is enough information from the four views to make this determination.

12

Below are 24 matchsticks arranged in a 3 × 3 grid. The circles represent the matchsticks that would have to be removed so that there are no squares of any size remaining in the grid. The minimum number, as you can see, is six matchsticks that need to be removed. There are several ways to accomplish this.

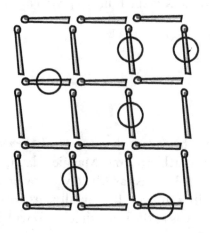

Now, what is the minimum number of matchsticks that need to be removed in a 4 × 4 grid of 40 matchsticks? Remember, there can be no squares, of any size, remaining.

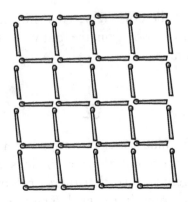

13

For what value K is the following system consistent?
1) P + Q = 6
2) KP + Q = 18
3) P + KQ = 30

14

Molly's yard has twice the area as Mickey's yard and 6 times as much area as Maggie's lawn. Maggie's mower cuts 1/3 as fast as Mickey's mower and 1/5 as fast as Molly's mower. If they all start to mow their respective lawns at the same time, who will finish last?

15

A factory makes gears for heavy machinery. Because of the intense heat and friction, a solid form of resin sticks is manually inserted into one of the machines at regular intervals to aid the grinding process.

Each resin stick has enough left over residue so that four pieces of it can be reformed to make another whole new piece of resin. At the beginning of the second shift, a worker notices 24 individual left over residue pieces. How many new resin sticks will these 24 make?

16

SEND
+MORE   D = 7, E = 5
MONEY

The well-known puzzle above first appeared in 1924 and is known as a cryptarithm or alphametic. Each letter has a positive integer value and for most alphametics, the range of values is between 0 and 9. No word can begin with a zero but that's about the only restriction. Often times, some of the values of letters are given to the solvers as hints. This also helps in keeping the answer to a single solution. Many alphametics have multiple solutions unless values are stipulated.

This particular puzzle has one solution only … and if you'd like to try to solve it, I'll give you two clues:

D = 7 and E = 5. There are eight different letters used, which means only eight numbers somewhere between 0 and 9 will be used.

The lead-in with this puzzle is usually stated as something to the effect that this was a note from a college student to his parents asking for a specific amount of money to cover tuition ... but that was a long time ago, so it's time for an update. Here's the new revised version:

SEND
MORE
LOTS
+MORE
GREEN

Let D = 7, E = 1, and O = 6

Now, see how long it takes you to come up with both solutions ... share them with family and friends. These puzzles can be great fun.

17

You have two pieces of string of varying thickness that each burn for 40 seconds. You want to measure 30 seconds. You can't cut either string in half because the fuses have different thicknesses and you can't be sure how long each will burn.

How can you accomplish this without bending, folding, or cutting the strings?

18

Let's say you wanted to calculate the number of license plates that could be issued by using three letters and three numeric characters (e.g., AAA123), using all letters of the alphabet and numbers 000 thru 999.

1) Assume that any of the letters or numbers can be in any of the six positions, that is: AAA000 or 000AAA or A0A0A0 or A1B2C3.
2) You cannot create plates that have all numbers (123456) or all letters (ABCDEF). Each plate must have three numbers and three letters.

How many possibilities for plates are there this way?

19

You are in a dark room sitting at a table. On a speaker, a voice tells you there are 26 nickels on the table – 10 are heads and 16 are tails. You are then told you will win $1,000 if you can separate the coins into two groups with the same number of coins heads up in each group, and you have one minute to accomplish this in total darkness. Could you do it?

20

If the stack of cubes below was originally 4 × 4 × 4, is A, B, C, D, or E the missing piece from the broken cube? Note1 : All rows and columns run to completion unless you actually see them end. Note 2: The missing piece is to be inserted upside down to complete the cube.

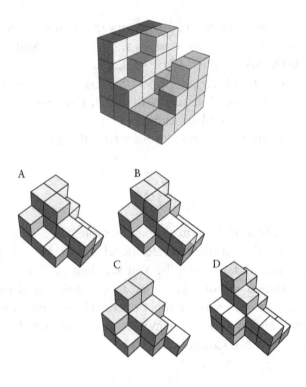

21

When a certain positive whole number (N) is divided by x, the remainder is 7. If the original number (N) is multiplied by 5 and divided by x, the remainder is 10. What is the least possible value of N?

22

I once wrote a puzzle where I asked the solver to use four different weights of counterbalance to weigh objects on a scale. I told them the weights have to be able to balance any whole number weight from 1 to 40 ounces. While I stated they could use four weights only, it is readily apparent that they will need more than one each of some of the weights. I asked what that minimum number was and how many of each of the weights would meet the requirements. It turns out that using two weights each of 1, 3, and 9 ounces plus one weight of 27 ounces will work (for a total of seven weights).

Here's the new puzzle. Is it possible there is an alternative solution(s) using four different weights? Again, some of the weights will have to be used more than once but seven weights total can't be exceeded.

23

Bella is doing a lab experiment and just realized she has put a certain solution in the wrong cylindrical jar. She needs another cylindrical jar with 30% larger diameter but the same volume as the jar she's currently using. If the diameter of the new jar is increased by 30% without the volume changing, by what percent must the height be decreased? Remember, the volume of a cylinder is $V = \pi R^2 h$.

24

The zeros that end a given number are called "terminal zeros." For example, 830,000 has four terminal zeros and 803,000 has three terminal zeros. How many terminal zeros are in the number represented by 15!?

25

One of the figures below doesn't belong with the others. The other four figures can be drawn without lifting a pencil, retracing or crossing any lines. Which is the odd one out?

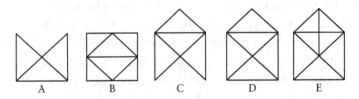

26

Can you create the following figure without lifting your pencil, retracting, or crossing another line? You can create the lines by going from vertex to vertex or line segments.

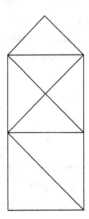

27

How many squares can you create in this figure by connecting any four dots (the corners of a square must lie upon a grid dot).

28

Two friends wish to meet for breakfast twice a month throughout the year but never on consecutive days. In how many ways can they choose those two days of the month so they never meet on consecutive days? (Consider a month to have 30 days.)

29

Triangle ABC is a right triangle with an altitude drawn from the vertex of the right angle to its hypotenuse.

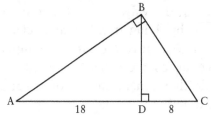

What is the length of AB? AD = 18, DC = 8

30

A certain city has done a thorough investigation and found that 2% of its citizens use drugs. That figure is solid and not in dispute. Suppose a drug test is employed that is 98% accurate. This means if a citizen is a user, the test will be positive 98% of the time. If the individual is a non-user, the test will be negative 98% of the time.

A person is selected at random, given the test, and the results are positive. What is the chance the person is a drug user?

31

On the planet Xenon there is only one airport and it is located on its North Pole. There are only three airplanes on the planet. Each plane's fuel tank holds just enough fuel to allow each plane to make it one way to the South Pole. There is an unlimited supply of fuel at the airport and the airplanes can transfer their fuel to one another. Your mission is to fly around the globe/planet with at least one airplane covering the circular journey and flying over the South Pole on its return to the North Pole. The planes may stop at any time along the way. (Note: the journey must be accomplished by flying in a "great circle." All great circles of a sphere have the same circumference and the same center as the sphere. A great circle is the largest circle that can be drawn on a sphere.) How can this be done?

32

Imagine you have a square piece of ordinary paper. Can you tell me a way in which you can divide this square into thirds, using a pencil and straightedge only? No compasses, protractors, or any measuring devices. You may fold the square once.

33

You are standing outside a 30-story building holding two identical glass spheres. You are told that either sphere, if dropped from the roof, would shatter upon hitting the earth, but that it would not necessarily break if dropped from the first story. Your task is to identify the lowest possible floor from which you can drop a ball and break it.

In the general case, what is the smallest number of drops required to guarantee that you have identified the lowest story? NOTES: 1) Both balls have the same minimum breakage story; 2) You only have two balls to use. If one breaks, it cannot be used for the rest of the experiment.

34

A mathematician needs to get through a train tunnel by foot. She starts through the tunnel and when she gets 1/3 the way through the tunnel, she hears a train whistle behind her. She begins to wonder how much faster the train is going than she is. At this point, we don't know how far away the train is, or how fast it is going (or how fast she is going). Here's what is known:

1) If she turns around and runs back the way she came, she will just barely make it out of the tunnel alive before the train hits her.

2) If she keeps running through the tunnel, she will also just barely make it out of the tunnel alive before the train hits her.

Assume she runs the same speed regardless of which path she chooses and reaches top speed instantaneously. Assume the train misses her by an infinitesimal amount (so that there are no extenuating circumstances that would make the puzzle unsolvable or a non-puzzle. In other words, don't make a case that she shouldn't be in the tunnel in the first place or that she still might get squished or anything that would cancel a solution.)

How fast is the train going compared to the mathematician?

35

If the stack of cubes below was originally 5 × 5 × 5, is A, B, C, D, or E the missing piece from the broken cube? Note 1: All rows and columns run to completion unless you actually see them end. Note 2: The missing piece is to be inserted upside down to complete the cube.

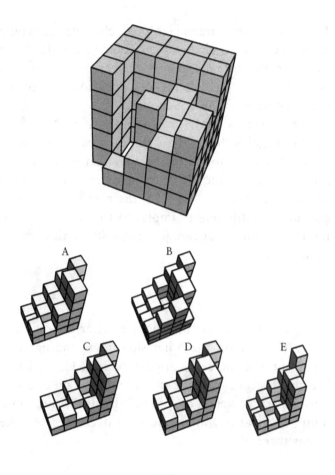

36

There is a fun old puzzle that asks the following:

At one point, a remote island's population of chameleons was divided as follows:

13 red chameleons

15 green chameleons

17 blue chameleons

Each time two different colored chameleons would meet, they would change their color to the third one (i.e. If green meets red, they both change their color to blue.) Is it ever possible for all chameleons to become the same color? Why or why not?

The answer is "no" – it is not possible. One of the most interesting things about this puzzle is that most of the explanations found online are either wrong or are not clear...and none I have found offer an example of when the answer is "yes." So, the new puzzle is this: give me two different examples (with different sets of numbers/colors) that can give chameleons all with the same color.

37

There are four cards that have a letter on one side and a number on the other side. Someone has laid them out and the cards appear as 2, 7, G, A. There is a rule that a card with an odd number on one side must have a vowel on the other. What is the minimum number of cards you should turn over to prove the rule is true?

38

You die and the devil says he'll let you go to heaven if you beat him in a game. He sits you down at a round table. He divides a large pile of quarters in two so you both have the same amount of quarters in your own pile. He says, "Ok, we'll take turns putting quarters down, no overlapping allowed, and the quarters must rest on the table surface. The first guy who can't put a quarter down loses. You cannot shift or try to squeeze any quarter into a space that moves another quarter." The devil says he wants to go first.

You realize that if the devil goes first, he probably has a strategy to win. You haven't had time to think this through yet, so you ask for some time to consider your options. He grants you 15 minutes. At the end of that time you know how to beat him...but you also know you must go first for your strategy to win. You convince him to let you go first, saying you really didn't have enough time to consider a strategy so you should at least go first. What is your winning strategy?

39

There's an old puzzle that asks, "If a hen and a half lays an egg and a half in a day and a half, how many hens will it take to lay 6 eggs in 6 days? The answer is 1.5 hens. Here's the new puzzle: at this rate, how many eggs will one hen lay in one day?

40

Imagine there are three boxes sitting on a shelf. You cannot see what is in them. You know one of the boxes has two bags of cherry gumdrops, one of the boxes has two bags of orange gumdrops, and the third box has one bag of cherry and one bag of orange gumdrops. You reach in one of the boxes and pick a bag of cherry gumdrops. What is the probability the second bag of candy in that box is a bag of cherry gumdrops?

41

When the proper weights are assigned, this mobile is perfectly balanced. What are the three missing weights?

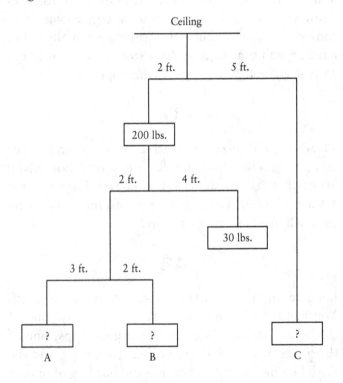

42

Tiki and Miki are both avid archers and have the exact same success rate of hitting the center of the target. That rate is 50%. They are going to enter a competition where they alternate shots until one hits the center of the target. Tiki will go first and then if necessary, third, fifth, etc. Miki will go second, forth, sixth, etc., if necessary. Who has the advantage, if there is one, and how much of an advantage does she have?

43

A mathematician likes to go to malls to walk up escalators at different speeds and leap multiple steps. She then creates puzzles for her classes where the students are required to figure out the number of steps in each escalator. On her latest adventure, she walked up an escalator (the escalator was moving upward) at the rate of one step per second and she covered 32 steps to bring her to the top. The next day she bounded three steps at a time and covered 48 steps to reach the top. It took her half the time as the first day. How many steps are in the escalator?

44

3/10 of all men in a certain group are married. 2/5 of all women in the same group are married. What fraction of the group is single men?

45

Four circles with a radius of 2 are each tangent to two sides of a square and externally tangent to a circle with a radius of 4.

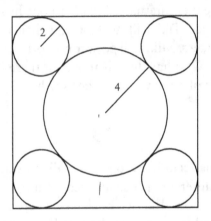

What is the area of the square?

46

Using three 2s and any math symbols and/or operations you choose, can you create 25?

47

A softball infield consists of a first baseman, second baseman, shortstop, and a third baseman. They are Brian, Rich, Kendall, and Steve, in some order but not necessarily in the order listed above.

1) Brian is younger than Kendall.
2) The first baseman is not related to any of the other infielders.
3) The second baseman and shortstop are brothers.
4) Kendall is Rich's nephew.
5) The second baseman is not the third baseman's uncle. And the third baseman is not the shortstop's uncle.
6) A father and son are part of this infield.

What position does each man play and how are they related?

48

A 200-pound man jumps off an 800-pound raft, travels 16 feet horizontally, and lands in the water. What maximum distance separates the man and the raft when he hits the water? The man and the raft have constant velocity. Disregard wind, friction, etc.

49

Moosaca is a mathematician who likes to create escalator puzzles by riding different escalators around New York. If she runs up 12 steps of the escalator, it takes her 60 seconds after that to reach the top of the escalator. If she runs up 18 steps, it takes her only 45 seconds after that to reach to top.

How many seconds would it take her to reach the top if she did not run up any steps? How many steps are there that are visible?

50

Mookie and Molly were running down an escalator (the stairs of the escalator were moving down also.) By the time Molly is ready to step off the escalator she has covered 60 stairs, while Mookie has covered 30 stairs. Molly's rate is faster than Mookie's. She runs four times as fast as Mookie. So, if they were on land, Mookie would take 40 steps to Mookie's 10.

How many of the escalator stairs are visible?

51

Martha, Mike, and Moosaca are going to play a series of one-set tennis single matches against each other. The winner of each set stays on the court and plays the player who had been idle. The loser of each set sits out the next match. At the end of the week, Martha played 15 sets, Mike played 14 sets, and Moosaca played 9 sets.

Who played the eleventh set?

52

Two friends decided to have a foot race. Each accelerated at a uniform rate from a standing start. Abbie covered the last 1/5 distance in 4 seconds. Bicky covered the last 1/4 distance in 5 seconds. Who won?

53

Using any math symbols, signs, or operations, complete the following to make the equations valid. You may not use any additional numbers. For example, you cannot use exponents, or radicals with a number indicating the root.

$$
\begin{array}{rccccc}
1) & 0 & 0 & 0 & = & 6 \\
2) & 1 & 1 & 1 & = & 6 \\
3) & 2 & 2 & 2 & = & 6 \\
4) & 3 & 3 & 3 & = & 6 \\
5) & 4 & 4 & 4 & = & 6 \\
6) & 5 & 5 & 5 & = & 6 \\
7) & 6 & 6 & 6 & = & 6 \\
8) & 7 & 7 & 7 & = & 6 \\
9) & 8 & 8 & 8 & = & 6 \\
10) & 9 & 9 & 9 & = & 6 \\
\end{array}
$$

54

Below is a figure built by using different sized square sheets of paper. What is the minimum number of square sheets needed to construct it?

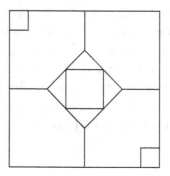

55

At a luncheon for physicians, all but 50 were neuro-surgeons. All but 60 were pediatricians and all but 70 were cardiologists.

How many of each group attended the luncheon?

56

At a certain school, the ratio of first graders to third graders is 4 to 7. The ratio of fourth graders to first graders is 3 to 4. The ratio of second graders to third graders is 6 to 5. What is the ratio of fourth graders to second graders?

57

How many numbers less than 1,000,000 have digits that total 3? Examples; 1,200, 111,000, 21,300.

58

You are arranging races for 25 horses on a track that can accommodate five horses at a time. Each horse always runs the distance in the same time and the horses have distinct speeds. You have no stopwatch, but can make deductions from the finishing order in the races. What is the smallest number of races needed to determine the three fastest horses, in order?

59

Your friend, Molly, was going through her attic and found an old Hula-Hoop. A Hula-Hoop is a toy that was popular years ago. It is made of plastic and is circular. The object of the toy or game was to see how long you could swing the hoop around your waist, not letting it fall to the ground. Here is the puzzle: Molly found that the circumference of the Hula-Hoop was 56 inches. Her waist is 28 inches around. She asks this question. If the Hula-Hoop makes contact with her waist at one point, how far will that point travel if the hoop makes exactly one revolution around Molly's waist? Keep in mind the point returns to the point from which it started right before Molly let the hoop swing around her waist one time.

60

At a major league tryout camp, 1/3 of all the pitchers are left-handed and 1/4 of the lefties are pitchers. If there are 240 participants at the camp, how many pitchers are present who are not left-handed? For the purposes of this puzzle assume that all non-pitchers are left-handed.

61

Solve for x:

$$\frac{1}{27} \cdot 3^{59} \cdot \frac{1}{243} \cdot 27^x = \frac{1}{3} \cdot 3^x$$

62

Marty is doing a biology experiment where he is required to log the growth rate of a certain animal species. He was given a male and female to begin the experiment. Each pair will begin to breed at the age of two months. Once a pair is old enough to breed, it gives birth to a new pair each month.

How many pairs will there be at the end of one year, including the original pair? Note: Count the two months as if the original parents were put together on January 1 and had their first offspring.

63

What is the greatest prime factor of $4^{19} - 2^{34}$?
A) 2   B) 3   C) 5   D) 7   E) 11

64

The dreaded cube-eaters from the fourth dimension descend upon a stack of 27 identical sugar cubes. Cube-eaters can only eat to the center of a cube. When they reach the center, they always make a 90° turn and proceed to the next cube. They never reenter a cube. If a cube-eater enters at location A, what is the minimum number of cubes it will eat through to reach the cube at location B?

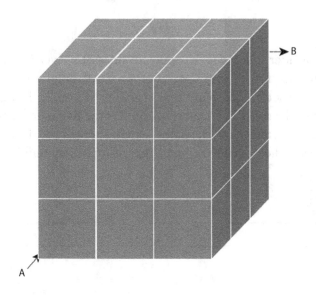

65

What is the smallest positive integer K such that the product $396 \times K$ is a perfect square?
A) 5   B) 7   C) 9   D) 11

66

A bicycle is three times as old as its tires were when the bicycle was as old as the tires are now. What is the ratio of the tires' current age to the bicycle's current age?

67

Consider two cities on opposite sides of a river (parallel lines for shores). You are the Chief Engineer of a construction company that is going to build a bridge over the river. Your job is to connect the two cities with the shortest possible path, which crosses the bridge perpendicular to the river. Where should you build the bridge?

City 1 •

• City 2

68

The following puzzle has been floating around at least 50 years (author unknown):

If Frank and Sam total their ages, the sum is 49. Frank is twice as old as Sam was when Frank was as old as Sam is now. How old are Frank and Sam?

You can find numerous references to this puzzle, with correct solutions, all over Puzzleland, especially online. However, in all the years this puzzle has been around, I have never seen a reasonable mathematical explanation expressed in algebraic form or otherwise. The answers to their ages are not difficult to guess through trial and error. But my puzzle is this. Can you show a complete solution set and defend it mathematically? I'll give you my solution in the answer section, but there may be several ways to approach this.

69

Figures 1, 2, and 3 are squares. Square 1 is 7x7; square 2 is 9x9. Points A, B, and C are collinear. What is X?

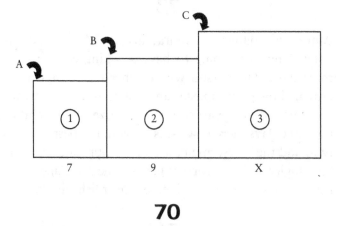

70

In a certain manufacturing plant, a new machine has been working only two days out of every three. What is the probability that this machine will work at least four out of five specified days in the future, all things remaining the same as they are now?

71

If $x + \dfrac{1}{y} = 12$ and $y + \dfrac{1}{x} = \dfrac{3}{8}$, what is the smallest value of $x \times y$?

72

A container holds an equal number of nuts and bolts. The mass of the bolts is twice the mass of the nuts. When half the nuts are removed, the mass of the contents is 90g. What will be the mass of the contents when half the bolts are then removed?

73

Amiro, his daughter Amira, his son Amir, and his sister Amire are going to have a fishing contest over three days. The person who catches the most will be crowned the grand master and the person who catches the least will be awarded a booby prize. At the end of the three days, here is what is known: the best angler's twin and the worst angler are of the opposite sex. The best angler and the worst angler are the same age. Who is the best angler (determined by most fish caught)?

74

The Delta-Turnaround is a commuter train system where Park Street is the central arrival and departure point. The trains run from Miller Square to Murphy Center 24 hours a day. Both Miller Square and Murphy Center have their own locomotives in their respective terminals to make the turnaround. Both end terminals have fixed schedules.

Miller
Square

Park Street

Murphy
Center

Molly arrives at Park Street daily at random times. Nine days out of 10 she notices that the first train that arrives is headed to Miller Square. How can this be? Intuition would tell you that over time, the number of arrivals and departures of the trains from each terminal would be equal. Can you give a reason why this isn't so?

75

Two rockets are launched simultaneously from two different positions.

Rocket A will land at the same spot from which Rocket B was launched, and Rocket B will land at the same spot where Rocket A was launched, allowing a small distance to the left or right to avoid a midair collision.

The rockets are launched from the same angle, and therefore travel the same distance both vertically and horizontally. If the rockets reach their destinations in one and nine minutes, respectively, after passing each other, how much faster is one rocket than the other?

A   B

76

1) Multiply 50 × 50 and raise that answer to the 1000th power. 2) Multiply 100 by 100 and raise that to the 500th power. How many times would you have to multiply the result of 2) to get 1)?
a) 50 to the 10th power
b) 25 × 25 to the 500th power
c) 50 × 50 to the 1000th power
d) 1,000,000 squared

77

Molly drives to work 42 miles. Because of construction, she can average only 12 miles per hour for the first 21 miles of the drive. How fast must she go on the second half of her trip to average 20 miles per hour for the entire trip?

78

A chemical distribution company sells rare compounds to research labs. One such compound may be purchased at $30 for 1/60 oz. or $60 for 1/30 oz. Many of the labs complained that 1/60 oz. was too little and 1/30 oz. was too much of the compound. An enterprising start-up company then came up with 1/45 oz. of the same compound for $45. Unfortunately, this company was forced out of business after one year. Can you offer any guesses as to why?

79

Below is an engine and four boxcars heading toward switch C. The configuration of trackage, ABC, is called a wye. Its purpose is to allow engines and boxcars to be turned to head in different directions. When the engine and four boxcars leave town, they will depart southwest through the B switch. The train needs to look like this when it leaves town:

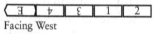

Facing West

How can this be accomplished if the train has to enter segment BC first? (There is ample switching room for the engine and four boxcars beyond each switch.)

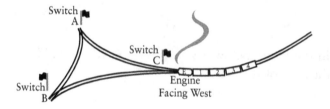

80

Eddie Fontani, head of the local racetrack, was shot and killed in a gangland killing. The police brought in five suspects for questioning. Each made three statements but the police knew that for each person, two of the statements were true and one was false.

Louie: I didn't kill Fontani. I have never owned a .38. Sawed-off Pete did it.

Robbie: I didn't shoot Fontani. I never owned a pistol. The others are all passing the blame around.

Doogie: I am innocent. I never saw Benny before. Sawed-off Pete is guilty.

Sawed-off Pete: I didn't do nothin' to nobody. Benny is the guilty one. Louie did not tell the truth when he said I did it.

Benny: I had nothin' to do wit' Fontani's murder. Robbie is the one who did it. Doogie and I are old pals.

Who killed Fontani?

81

In the figure below, PS = 8; PQ = 6; RS = 18. What is the area of Δ PTR?

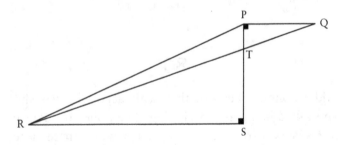

82

Below are three identical dice. However, these are not standard dice. The opposite faces do not total 7. Each of the numbers from 1 to 6 appears on exactly one face of each die. The touching faces of the dice have the same number.

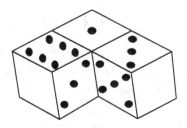

Fill in the net with the appropriate number (representing the dots) found on each face.

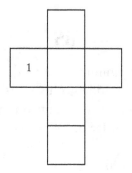

83

a) What fraction in $Base_{10}$ is equal to .132 $Base_4$?

b) The fraction 1/9 is equal to the repeating decimal .1111... What is 1/9 in $Base_{12}$ decimal form?

c) 11.11 in $Base_2$ is equivalent to __?__ in $Base_{10}$?

d) What is the number 1,357.975 (Base$_{10}$) when it is converted to Base$_{1/10}$ (Base one tenth)?

84

In how many discrete ways can four cubes be placed together face-to-face? (No mirror images and no partial coverage of any face.) For example:

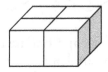

85

Below is the beginning of an equation that needs your attention. By adding plus signs and minus signs between or in front of the digits on the left side (without changing the order of the digits), the equation can be made correct.

123456789 = 100

How many different ways are there to make the equation correct?

86

If it takes 30 seconds to blow up a balloon to 1 foot in diameter, how much longer will it take to inflate the balloon to 3 feet in diameter? Several things to know:

1) the pressure the balloon exerts on the air inside is proportional to the surface area of the balloon, 2) the balloon retains its shape from start to finish, and 3) the air going into the balloon is entering at the same rate throughout.

87

Which figure is different from the others in the way that it is oriented?

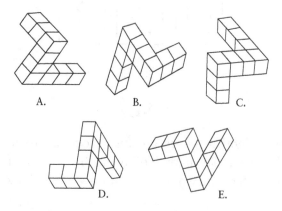

A.   B.   C.

D.   E.

88

In the land of Nonnon, the number system resembles ours in the symbols they use for the numbers, but the results of their math operations are different. For instance, $6 \cdot 4 = 26, 6 \cdot 5 = 33, 5 \cdot 7 = 38$. What does $4 \cdot 5 \cdot 6$ equal in Nonnon's number system?

89

Imagine a square sheet of paper. Fold it diagonally to form a triangle. Then fold it again in half so that the two equal sides meet tip-to-tip. Which shape will it look like?

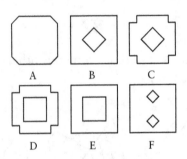

90

Several identical cubes are fused together to form a solid object. Given the following five views of that object, determine what the sixth view would look like.

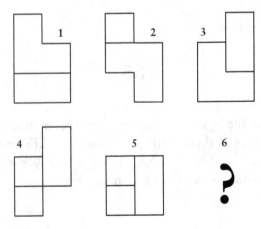

91

Nine identical sheets of paper are used to create the design below. If "D" is placed first, "F" placed seventh, and "I" placed last, in what place would you find "B"? Where would you find "H"?

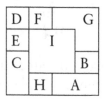

92

The probability that any one person chosen at random being born on a Friday is 1 in 7 or 14%. What is the probability that of seven people chosen at random that one or more was born on a Friday? What is the probability that exactly one person was born on a Friday?

93

Sometimes in school or business we are given information that looks impossible to decipher, only to find that applying a little "elbow grease" aids in sorting things out. Below are several statements that form a relationship between the letters A, B, C, and D and the numbers 1, 2, 3, and 4. Using the information, see if you can straighten out my confusion and identify each letter with its number.

If A is not 2 or 4, then B is 1.
If A is 3, then B is not 4.
If B is not 1, then D is 4.
If C is 3, then D is not 2.
If C is not 2, then D is 2.
If D is not 1, then A is 3.

94

How many individual cubes are in the stack below? All rows and columns run to completion unless you actually see them end.

95

The rabbits of Lepusville were in distress. They had lost too many of their fellow citizens to the sly 'ol fox. He was outsmarting them every time. They needed a plan so they could be safe when they left their underground homes. They decided to bring in the genius professor of mathematics at the University of Conejo, Professor Edwin T. Harvey. They explained their situation to Dr. Harvey and then told him these facts.

1) Through a combination of sight, sound, and smell every rabbit can detect a fox at 48 yards.

2) When a rabbit detects a fox, it runs immediately straight to its hole.

3) It was well known that a fox can run exactly twice as fast as any rabbit.

4) If a rabbit and fox reach the hole at the exact same instant, the rabbit is always the winner and survives the chase.

5) The diameter of the hole is considered negligible for the purpose of computing distance.

The community of Lepusville needed a sure-fire method for them to know the maximum distance from a hole that would guarantee a rabbit's safety going and coming.

The wise professor thought for a few minutes and then gave them their answer. They were shocked at the simplicity of the solution and there was not doubt in anyone's mind that it was correct and would protect them from that day forward.

A few months later Mr. Fox had to move his headquarters ... his days of catching any rabbit from Lepusville were over and he knew he had met his match.

What was the distance the professor said would guarantee each rabbit's safety?

96

Below are blocks – all the same size. Your job is to count how many times the sides or faces of a block touches the sides or faces of other blocks. Blocks that connect only at the edges or corners do not count.

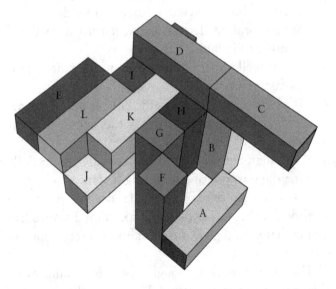

97

Here are several sequence puzzles where you are to find the missing number. But don't let the word "number" mislead you. Just because it contains a number does not necessarily mean the sequence is mathematical. Anything is fair game in sequences. No higher mathematics is needed, just a flexible, clever mind.

a) 1   4   2   8   5   7   _?_
b) 31   41   59   26   53   58   97   _?_
c) -40 -14 4   14   16   10   _?_
d) 5   13   10   16   _?_ 9   6   14
e) 1   2   4   12   24   52   _?_
f) 16   21   26   26   12   5   0   6   21   _?_

98

Using the weights of 10, 20, 30, 40, 50, 60, and 70, assign them once and only once to their respective boxes below so the mobile will balance. Each hash mark represents 10 ft.

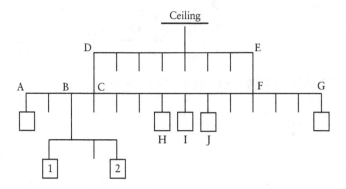

99

Using 4 twos and the math operations of +, −, ×, and ÷, what is the largest number that can be created? You may use the twos in exponents but no other math operations or symbols can be used.

100

What is the missing figure in this analogy puzzle?

```
B    B    C              D D D
B    B    C              D    D
B B B  •  C      • •  D D D  •  ?
B    B  •  C      • •              •
B    B    C C C C        D
                         D
```

101

Here's a new type of tic-tac-toe. We'll call it cit-cat-cot, because in this version, three in a row loses! Below is the start of a game and it's X's turn to move.

X		
O		O
O	X	

Where is the best square to place your X?

SOLUTIONS

1

Remove one student from each bench, including the one with four students to make a new bench of eight students. There must be seven full benches in order to do this. Therefore, there were seven benches of nine students plus the four for 67 students.

Following that there are some other ways to this. First:

Let n = Number of benches needed when seating eight per bench

m = Number of benches needed when seating nine per bench

s = Number of students

Writing the two equations for the numbers of students in terms of the seating on the benches, we have

$s = 8n + 3$

and $s = 9m + 4.$

Recognizing that the number of students is the same, we set the quantities equal:

$8n + 3 = 9m + 4.$

Then, noticing that n must be greater than m, we condense, to write

$8n - 9m = 1.$

Pulling out a common factor, we have

$8(n - m) - m = 1.$

Then, noticing that the quantity $(n - m)$ is actually the difference in the number of benches used to seat the students (between the 8-per-bench and 9-per-bench approaches), we define the difference as

$d = n - m.$

Then, substituting the difference d and rearranging, we have

$m = 8d - 1.$

Substituting back in one of the original equations yields

$s = 9(8d - 1) + 4,$ or

$s = 72d - 5.$

Since the number of students in a class must be a positive number, we see by inspection that d must be greater than 0. So, setting d to 1, to get the minimum number of students satisfying the equation, we have

$s = 72(1) - 5 = 67$

Minimum number of students = 67

Here is another way to solve this:

We know that $3 + 8x = 4 + 9y$

subtract 8y from both sides

$3 + 8(x - y) = 4 + y$

The smallest value of $(x - y)$ is 1

(since x is the = number of benches with eight students per bench and y is the number with nine per bench, so $x > y$ because it will take fewer benches to seat more students)

So the smallest value is when $(x - y) = 1$ plug in the equation

$3 + 8 = 4 + y$

so $y = 7$.

That's it!

$x + 3 = y + 4$

or $x - y = 1$

which, according to what I said above, is the smallest positive difference in the number of benches!

2

5/8. When you switch the marbles to their new boxes, you have the probability of having

W and BW or

B and WW

In the four cases,

A. W   **B.** BW

C. B   **D.** WW,

1/4 the time you get a white marble by choosing **A**. 1/4 the time you get a white marble from choosing **D**. In the case of choosing **B**, you get a white marble 1/2 of 1/4 or 1/8 the time.

Total probability of a white marble: 1/4 + 1/4 + 1/8 = 5/8.

3

The watermelons weigh 100 kg at the end of the day. In the morning, the 200 kg of watermelons are 99% water. So the non-water part of the watermelons has a mass of 2 kg. At the end of the day, the melons are 98% water. The remaining 2% is still the 2 kg of non-water material (which does not change when the water evaporates). If 2% equals 2 kg, then 100% equals 100 kg. So, Molly has 100 kg of watermelons left at the end of the day.

4

From lowest to highest value:

1) $\ln n$
2) n
3) n^c
4) c^n
5) $n!$
6) n^n

5

−5. One way to solve this is to examine the difference between the numbers to see if a pattern develops.

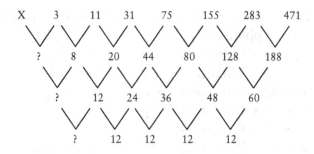

Once you see a pattern, then work your way back to the top to find the first number. Keep in mind when finding the differences between numbers, you have to take into account if the difference is negative or positive.

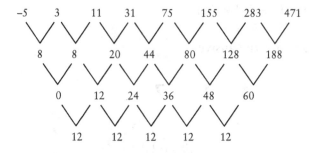

6

If 3 and 5 divide the number, then 15 must divide the number. So the number is a multiple of 15.

Now we need to look at the remainders when you divide multiples of 15 by 7.

Here are the first few multiples of 15 written as multiples of 7 plus a remainder:

15 = 14 + 1 (remainder = 1)
30 = 28 + 2 (remainder = 2)
45 = 42 + 3 (remainder = 3)
60 = 56 + 4 (remainder = 4)
75 = 70 + 5 (remainder = 5)

So the smallest multiple of 15 with a remainder of 5 when divided by 7 is 75.

Start with the smallest multiple of 15, namely 15, and divide by 7 you get the following equation (note: the remainder 1):

15 = 2 × 7 + 1

We want a remainder of 5, so multiply both sides of this equation by 5:

5 × 15 = 5(2 × 7) + 5
or 75 = 70 + 5

So 75 is the answer.

7

```
              3091
        33) 102003
            99
            300
            297
              33
              33
               0
```

If you spend a little time going over the puzzle, you see that the 3 in the dividend needs to be brought down alongside the bottom 3 to make 33. Then 33 is divisible by 33 one time.

You also know the first three numbers of the dividend must be 102. Thirty-three goes into some 3-digit number and has a remainder of 3, on the third line. Thirty-three goes into 102 three times with a remainder of 3. From there, it is easy to fill in the remaining numbers.

8

10 units.

Area = 192, Let x = length of BE

AE = 6 units.

$192 = 1/2 \cdot 6x + 21x$

$192 = 3x + 21x$

$192 = 24x$

$x = 8$

$C^2 = A^2 + B^2$

$(AB)^2 = 6^2 + 8^2$

$(AB)^2 = 100$

AB = 10

9

45. Since the median is 9, the three numbers are x, 9, and y.

$1/3\,(x + 9 + y) = x + 12$ and $1/3\,(x + 9 + y) + 18 = y$

Add the two equations:

$2/3\ (x + 9 + y) + 18 = x + 12 + y$

Multiply each side by 3:

$2x + 18 + 2y + 54 = 3x + 36 + 3y$

$72 = x + y + 36$

$x + y = 36$

Since the median is 9, the sum is $9 + 36$ or 45.

10

5 miles. distance = rate × time

d_{AL} = Distance Traveled by Alison

d_{AM} = Distance Traveled by Amelia

$d_{AM} = rt$

$d_{AL} = \left(\dfrac{3}{5} \cdot r\right)(3t) = \dfrac{9}{5}rt$

$(d_{AL} + d_{AM}) = 14 = \dfrac{9}{5}rt + \dfrac{5}{5}rt$

$14 = \dfrac{14}{5}rt$

$rt = d_{AM} = 5$ miles

11

"B" is incorrect. The correct view should be:

12

By visual inspection, it can be seen that 10 is the minimum number. Here's one way this can be accomplished:

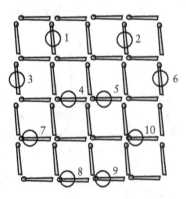

13

K = 7. The sum of the 2nd and 3rd equations is (P + Q) + K(P + Q) = 48

Divide each side by P + Q…

1 + K = 48/(P + Q)

1 + K = 48/6

1 + K = 8

K = 7

14

Molly will finish last. Here's one way to look at this:

Molly's Lawn Size = x

Mickey's Lawn Size = $x/2$

Maggie's Lawn Size = $x/6$
Molly's Mowing Speed = y
Mickey's Mowing Speed = $3/5y$
Maggie's Mowing Speed = $1/5y$
You have to compare their respective areas with their mowing speeds to see who is the slowest:

$$Molly = \frac{x}{y}$$

$$Mickey = \frac{\frac{x}{2}}{\frac{3}{5}y} = \frac{5}{6} \times \frac{x}{y}$$

$$Maggie = \frac{\frac{x}{6}}{\frac{1}{5}y} = \frac{5}{6} \times \frac{x}{y}$$

You can see Molly is the slowest and Mickey and Maggie tie.

15

Eight new sticks. The worker can make six new sticks from the 24 left over pieces, which will generate six left over residue pieces. They, in turn, can be used to make one more new resin stick, with two pieces left over. After using the new stick, there will be two more pieces left over. Couple those two left over pieces with the previous two left over pieces and you can create one more whole stick, which makes a total of eight new sticks.

16

SEND	9567	
+MORE =	+1085	D = 7, E = 5, M = 1, N = 6,
MONEY	10652	O = 0, R = 8, S = 9, Y = 2

SEND	5147	D = 7	N = 4
MORE	9681	E = 1	O = 6
LOTS	3605	G = 2	R = 8
+MORE	+9681	L = 3	S = 5
GREEN	28114	M = 9	T = 0

17

Light both ends of one of the strings (it doesn't make any difference which one). It will burn down in 20 seconds regardless of how evenly it burns.

At the same instant, light one end of the second string. It will start to burn while the other is totally used up. In 20 seconds, when the first string is finished burning, light the second end of the second string. This will burn what's left in 10 more seconds – for a total of 30 seconds.

18

351,520,000.

There are 20 ways to pick the three positions for the letters. This is found by "6 choose 3" which is expressed as $\frac{6!}{3!3!} = \frac{6 \times 5 \times 4 \times 3 \times 2 \times 1}{3 \times 2 \times 1 \times 3 \times 2 \times 1} = 20.$

Secondly, there are 26^3 ways to distribute the letters followed by 1000 for the digits. That gives you $20 \times 26^3 \times 1000 = 351,520,000$.

19

You can accomplish this if you knew to randomly select 10 coins and put them off to the side as Group I and the remaining 16 coins to the right as Group II and then turn over all the coins in Group I. You will now have the same number of heads up coins in each group.

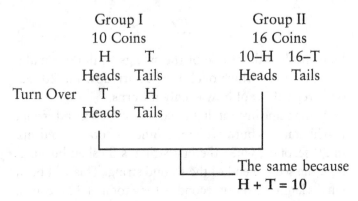

	Group I		Group II	
	10 Coins		16 Coins	
	H	T	10–H	16–T
	Heads	Tails	Heads	Tails
Turn Over	T	H		
	Heads	Tails		

The same because H + T = 10

20

Figure B is the missing piece from the broken cube.

21

N = 32

Let y = the integer quotient of N

$y = N/x$, forgetting the remainder. Then, $N = x \times y + 7$.
Now consider 5 times N: $N = 5(x \times y) + 35$.
$5[(x \times y + 7)] = 5(x \times y) + 35$

If you divide this by $x \to 5y + 35/x$. $35/x$ is the fraction that will determine the remainder of 10. If $35/d = 1$ remainder 10, then x is 25. 25 + 7, the remainder, gives us 32 as a solution.

$\frac{32}{25} = 1$ with a remainder of 7.

22

Here are three solutions. Did you find another?

2 - 16 oz weights
1 - 8 oz weight
1 - 4 oz weight
3 - 1 oz weights
—————
7 Total

or

1 - 1 oz weight
2 - 2 oz weights
1 - 5 oz weight
3 - 10 oz weights
—————
7 Total

or

1 - 1 oz weight
3 - 2 oz weights
1 - 7 oz weight
2 - 13 oz weights
—————
7 Total

23

59.1% reduction.

When the diameter is increased by 30%, that's the same as 13/10, so the area of the base is increased by $(13/10)^2 = 169/100$. To keep the volume the same, the height must be $\dfrac{1}{169/100} = \dfrac{100}{169} = 59.1\%$.

24

The number will end in three zeroes.

$15! = 1{,}307{,}674{,}368{,}000$. One way to look at this is to realize $15! = 15 \times 14 \times 13 \times 12 \times 11 \times 10 \times 9 \times 8 \times 7 \times 6 \times 5 \times 4 \times 3 \times 2 \times 1$. You have a 5×2 for one zero, a $10 \times$ any other number for another zero and $15 \times$ an even number to produce the third zero.

25

Figure E is the odd one out.

26

A and B are the odd vertices where the path must begin and end. Begin at A and move consecutively through the numbers.

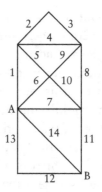

27

There are eleven squares total; five small, four medium, and two large.

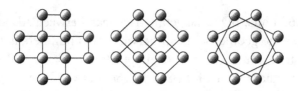

28

405 ways.

Let's say they meet on the 1st of the month, as one example. Then they could meet on the 3rd, 4th, 5th, 6th, ... 29th. They can't met on the 30th of this month because they will meet again on the following day, which is the 1st of the following month. So, for any month, this gives $27 \times 30 = 810$ choices.

However, this counts every selection twice (meeting on the 3rd and the 12th is the same as meeting on the 12th and the 3rd). So we need to divide 810 by 2 to come up with the correct answer – 405 choices.

If you were to view this puzzle as "non-cyclical" (i.e., the 30th of the month and first of the following month as not being two days together, then the answer would be 406. A mathematical way to show this is by finding "30 choose 2" which is 435. You then have to subtract all the days where the combination days in a month when they could not meet...(1, 2) (2, 3) (3, 4)... (29, 30) ... there are 29 such combinations in any month. 435 – 29 = 406.

29

21.63 units. BD is the geometric mean between AD and DC. The geometric mean between two numbers is the positive square root of their products. $18 \times 8 = 144$. The square root of 144 is 12. So BD = 12. Since we now know AD = 18 and BD = 12, we can find AB with the Pythagorean Theorem:

$C^2 = A^2 + B^2$.
$C^2 = 18^2 + 12^2$; $C^2 = 144 + 324$; $C^2 = 468$ $C = 21.63$

30

There is only a 50% chance the person is a drug user. Suppose there are 100,000 people in the city.

We assume that 98% of them (98,000) are non-users and 2% (2,000) are users. Of the 98,000, 98% of them (96,040) will test negative. That means 2% of them (1,960) will test positive. Of the 2,000 users, 98% of them (1,960) will test positive. That means 2% of them (39.2) will test negative. That means there are 1,960 non-users who will test positive (false positives) and 1,960 users who will test positive. So, the conclusion is when a person is found in the positive group, there's only a 50/50 chance he/she is a user.

However, if a person is randomly chosen and tests negative, the situation is entirely different. As we know, there are 39.2 "false negatives" and 96,040 true negatives – a total of 96,079.2. So there is more than a 99% chance that this person is *not* a drug user.

31

1) Think of dividing each half of the globe into thirds. From the planet's North Pole to its South Pole is three thirds and from the South Pole to the North Pole is three thirds.

Two airplanes start from the North Pole and at the 1/3 mark, one of the planes fuels the other and returns to the airport on the North Pole. The refueled plane continues on to the 2/3 mark. It has 2/3 of a full tank when it arrives at the 2/3 destination mark.

2) There are two airplanes now on the North Pole. They leave simultaneously and fly to the 1/3 mark where one of the planes fuels the other and returns to the airport on the North Pole. The other plane continues on to the 2/3 mark where it will fuel the plane already there. You now have two planes at the 2/3 mark and each has 2/3 of a tank of fuel – but not for long. One of the planes fills up the other with a full tank of fuel and this plane returns to the 1/3 mark, waiting to be fueled by the plane which will leave the airport to refuel it – then they will both return to the airport. The plane at the 2/3 mark with the full tank then continues the global journey – rounds the South Pole and lands on the other side of the globe at the 1/3 mark.

3) Now, all that needs to be done is to have one of the two planes at the airport fly to the other 1/3 mark to refuel the "roundtripper" plane now there. After refueling, they both fly home to the airport and the mission is accomplished.

32

Here is one way to divide a square into thirds:

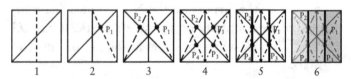

1) Fold the square in half as shown by dotted line. Now unfold the square back to the original size and draw a diagonal as shown.
2) Steps 2-6 are completed with a pencil and straight-edge.

33

To test this, start from floor number 8 and drop a ball. If it doesn't break then go to floor 15. If it doesn't break there go to floor 21. If it doesn't break there, go to floor 26. If it doesn't break there, go to floor 30. If it breaks there you have to test floors 27, 28, and 29. You have taken 8 drops to test this.

Now, take the other case: suppose you drop one ball on floor 8 and it breaks. You now have to test floors 1 through 7 to find out if that ball will break. That, also, totals 8 as does each testing stage between any of the two consecutive floors you've chosen. Notice how the difference between each set of consecutive floors decreases by 1...8, 15, 21, 26 and 30? An easy way to find the minimum number of drops for any floor is to add $1 + 2 + 3 + 4 + \ldots$ until the sum reaches the number of the top floor. As I demonstrated in the first sentence, once the number has been met (even if the last added number surpasses the top floor number when adding it, it is taken to that whole number to determine the minimum number of attempts) that is the solution: $1 + 2 + 3 + 4 + 5 + 6 + 7 + 8 = 36$ so the threshold of 30 has been passed after the 8 is added. Therefore, it takes a minimum of 8 drops.

34

The train is going three times as fast as the mathematician.

We know that the train and mathematician will reach the start of the tunnel at the same time if the mathematician turns around, in that time she travels 1/3 the length of the tunnel.

If she travels on towards the end of the tunnel she will be able to travel an additional 1/3 of the length of the tunnel before the train reaches the start of the tunnel, putting her at the 2/3 way point, (leaving one third of the tunnel until she reaches the end).

For both the mathematician and the train to reach the end of the tunnel at the same time, she has to travel 1/3 the length of the tunnel while the train has to travel the full length of the tunnel.

Therefore, the train must be traveling three times as fast as the mathematician.

35

Figure D is the missing piece from the broken cube.

36

Here are two ways this can work:
   Examples:

1) If any two of the beginning colors are the same number 15 red, 18 blue, 15 green...the reds and greens

meet to form 30 new blues and the total becomes 48 blues (30 + 18) or this situation:

2) Say you have 5 blue, 2 green and 1 red (8 total). Match one blue with one red, resulting in two new greens and leaving no reds. We now have 4 greens (we started with 2) and 4 leftover blues. Now match the 4 blues with the 4 greens and you will end up with 8 reds.

Any variations of these will, of course, be solutions, as long as they represent one of these two methods.

37

The minimum number of cards you should turn over is two and the cards are 7 (to see that the other side has a vowel) and G (to see if the other side has an odd number). You don't need to turn over A because the rule is that a card with an odd number must have a vowel but *not* that a card with a vowel must have an odd number.

38

Put the first quarter exactly in the center of the (circular) table.

Next, for each quarter the opponent places, place one directly opposite it. That is, place it so that the center of the table is halfway between your piece and the opponent's previous piece.

This will generate a completely symmetric layout of quarters on the table. This means that whenever the opponent selects a free space to place a quarter in, the space opposite is guaranteed to be free as well. Since you are always guaranteed an open space, you will never lose with this strategy.

39

The rate per hen per day is 2/3 egg. If a hen and a half lays an egg and a half in a day and a half, then one hen lays 1 egg in 1.5 days. 1.5 days is 3/2 days.

1) 1 hen × 3/2 days × rate per hen per day = 1 egg
2) 3/2 rate per hen per day = 1 egg
3) Multiply each side by 2/3...

Rate per hen per day = 2/3 egg

40

The answer is 2 out of 3 chances. Here is one way to look at this. There are 6 bags of candy.

Box 1: 1 Cherry 2 Cherry
Box 2: 3 Orange 4 Orange
Box 3: 5 Cherry 6 Orange

You picked a cherry bag, so you picked bag 1, 2, or 5. All the possibilities are equally likely, so the chances it was bag 1 or 2 are 2 out of 3.

41

A = 48 lbs.; B = 72 lbs.; C = 116 lbs.

Take a look at the 4 ft. segment supporting 30 lbs. 4 ft. × 30 lbs = 120 ft. lbs. If there are 120 ft. lbs. on the right side of this portion of the mobile then there must be 120 ft. lbs. on the left side – the side supported by 2 ft. We know 60 lbs. has to be distributed in a 2 to 3 ratio on the left side. 2 ft. × 60 lbs. = 120 ft. lbs. So we know it will balance with the right side. And since it must be in a 2 to 3 ratio, 48 lbs. will be the weight in A and 72 lbs. in B.

Now, we know this portion of the mobile balances because 3 ft. × 48 lbs. = 144 ft. lbs. = 2 ft. × 72 lbs. Now, add all the weights on the left side of the mobile – the weights supported by the 2 ft. segment at the very top. 200 + 30 + 60 = 290 lbs. 290 lbs. × 2 ft. = 580 ft. lbs. on the left side. We need the same on the right side supported by 5 ft. at the top. 5 ft. at the top. 5 times some number equals 580.

$5x = 580 \quad x = 116$

So, 116 lbs. is the weight of C.

42

Tiki will have a 2/3 chance of winning as opposed to 1/3 for Miki - so Tiki has twice the chance of winning over Miki. The math looks like this.

The first person's chances look like this:

1/2 (first try) + 1/2 × 1/2 × 1/2 (third try) + 1/2 × 1/2 × 1/2 × 1/2 × 1/2 (fifth try) + ... = 1/2 (1 + 1/4 + 1/16 + 1/64 + ...)

This computes to 1/2(4/3) or 2/3.

43

64 steps. Here is one way to think about this:

Let N = the number of steps

Let r = steps per second

$N - 32r = 32$

$N - 16r = 48$

Solving simultaneously,

$N - 32r = 32$

$-N + 16r = -48$

$-16r = -16$

$r = 1$ step per second. Plugging r back into either equation we find

$N = 64$ steps

44

40 percent.

Let M = number of men in the group and W = number of women.

$(3/10)M$ is the number of married men, so $(7/10)M$ is the number of single men.

$(2/5)W$ is the number of married women.

We want to compute the ratio $R = \dfrac{(7/10)M}{M+W}$.

Divide top and bottom of this fraction by M to get

$$R = \left(\frac{7}{10}\right)\left(1+\frac{W}{M}\right).$$

So, we just need the ratio of women to men. But we are given $(3/10)M = (2/5)W$ since those in the group who are married are married to people in the group.

So, $\dfrac{W}{M} = \dfrac{\dfrac{3}{10}}{\dfrac{2}{5}} = \dfrac{3}{4}$

So plugging this into equation R we get

$$R = \frac{\dfrac{7}{10}}{1+\dfrac{3}{4}} = \frac{2}{5}.$$

45

$88 + 48\sqrt{2}$ units.

One way to view this is to draw two radii from the larger circle to the centers of the two upper circles. This forms a right triangle of sides 6, 6, and $6\sqrt{2}$. The length of the hypotenuse of the right triangle plus twice the radius of a smaller circle is equal to the length of the side of a square: $6\sqrt{2} + 2 + 2 = (4 + 6\sqrt{2})$. Now square this for the area.

$(4 + 6\sqrt{2})^2 = 88 + 48\sqrt{2}.$

46

Here's one way:

$$\left(\sqrt{.2^{-2}}\right)^2 = 25.$$

47

Brian is the first baseman.
Kendall is the third baseman.
Rich is the shortstop.
Steve is the second baseman.
Rich and Steve are brothers.
Kendall is Rich's nephew and Steve's son.

Discussion:

Look at statement number 4. Rich and Kendall are related as uncle and nephew so neither is the first baseman. That means either Brian or Steve is the first baseman. Kendall has an uncle (Rich) and since Brian is younger than Kendall, the only relation he could be is Kendall's brother. That would make Kendall and Brian the second baseman and the shortstop (in some order) and Steve would be the first baseman and not related to anyone. That leaves Rich to be the third baseman and an uncle to Kendall. But that doesn't leave any room for the father. So, Brian couldn't be Kendall's brother. He is no relation and the first baseman. Since Kendall is Rich's nephew, that leaves Steve to be Kendall's father and a brother to Rich. Since the second baseman and shortstop are brothers, Kendall must be the third baseman. That leaves the brothers Rich and Steve to be the second baseman or shortstop, in some order. Look at statement number 5. Since the second baseman is not the third baseman's uncle, then the second baseman must be Steve, which means Rich is the shortstop.

48

If the man was standing above the waterline on the edge of the raft, he would have to travel downward as well as horizontally to get wet. It is the time for the downward fall that determines how long he is in flight. The time for the downward fall is the same as the time for his horizontal flight.

Given conservation of momentum for the jump, the raft initially would have (200/800) = 1/4 the velocity of the man in the opposite direction of the man, and therefore would have covered a maximum of 1/4 the distance. So the raft travels a maximum of 4 feet in the opposite direction the man travels. The maximum separation would be 16 feet + 4 feet = 20 feet.

49

It would take Moosaca 90 seconds to reach the top if she did not run up any steps. There are 36 visible steps.

Let the total number of steps in the escalator be x.

$(x - 12)/60 = (x - 18)/45$

$x = 36$. There are 36 steps.

$(36 - 12)/60 = (36 - 18)/45 = 2/5$ which is one step every 2.5 seconds. $36 \times 2.5 = 90$ seconds.

50

80 stairs. If the escalator were stationary, Molly would travel four times as fast as Mookie but Molly is only twice as fast, in this case, on the escalator. The speed of the escalator is the same for both and affects the ratio of their speeds as compared to land. When Molly has covered 60 steps, Mookie has covered 30. This means the stairs are moving twice as fast as Mookie. When Mookie actually takes 10 steps, the stairs add on their rate of 20 stairs covered and Mookie moves 30 stairs. Likewise, Molly has stepped 60 times and the stairs have moved 20 steps. So, 60 + 20 = 80 stairs.

Molly (x) + Stairs (y) = 2(Mookie (z) + Stairs (y))

$x + y = 2z + 2y$

$x = 4z$

$4z + y = 2z + 2y$

$y = 2z$, the stairs are moving twice as fast as Mookie.

$x + 2z = 2z + 2y$

$x = 2y$, the stairs are moving half the speed of Molly.

51

Martha and Mike played the eleventh set. The total number of sets played was
$$\frac{(15+14+9)}{2} = \frac{38}{2} = 19.$$
Moosaca played 9 sets and sat out 10 sets, which means he lost the even numbered sets, leaving Martha and Mike as the players of the eleventh set.

52

Bicky won. Abbie covered the last 4/20 of the race in 4 seconds; Bicky covered the last 5/20 of the race in 5 seconds. Both average 1/20 of the course per second for the respective distances. Since Bicky maintains this average over a greater distance, she finishes first.

53

Here is one set of solutions but there are several alternative ways to solve these:

1) 0 0 0 = 6 $(0! + 0! + 0!)!$ = 6

2) 1 1 1 = 6 $(1 + 1 + 1)!$ = 6

3) 2 2 2 = 6 $2 + 2 + 2$ = 6

4) 3 3 3 = 6 $(3 \times 3) - 3$ = 6

5) 4 4 4 = 6 $\sqrt{4} + \sqrt{4} + \sqrt{4}$ = 6

6) 5 5 5 = 6 $(5/5) + 5$ = 6

7) 6 6 6 = 6 $6 + 6 - 6$ = 6

8) 7 7 7 = 6 $7 - (7/7)$ = 6

9) 8 8 8 = 6 $\left[\sqrt{(8/8)+8}\right]!$ = 6

10) 9 9 9 = 6 $\left(\sqrt{9} \times \sqrt{9}\right) - \sqrt{9} = 6$

54

Seven sheets of paper are needed.

One large sheet that is placed first. Then, two squares 1/4 the size of the largest sheet. One of these is placed in the upper right quadrant; the second in the lower left quadrant. This gives the effect of having four equal-sized squares in the background. Now add the diamond square and the square inside the diamond for a total of five squares. Finally, the two small squares in the corners for a total of seven squares.

55

There were 90 physicians total at the luncheon. If 50 were not neurosurgeons, then that would be the total of the pediatricians and cardiologists. If 60 were not pediatricians, that would be the total of the neurosurgeons and pediatricians. If 70 were not cardiologists, then that would be the total of the neurosurgeons and pediatricians.

Each group is included twice in the three totals. So,

$50 + 60 + 70 = 180$

$180/2 = 90$

$90 - 50 = 40$ neurosurgeons

$90 - 30 = 30$ pediatricians

$90 - 20 = 20$ cardiologists

56

The ratio of fourth graders to second graders is 5 to 14. Here's one way to solve this:

$$\frac{1\text{st}}{3\text{rd}} = \frac{4}{7} = \frac{Fi}{7} \text{ So, } Fi = \frac{4}{7} \times T$$

$$\frac{4\text{th}}{1\text{st}} = \frac{3}{4} = \frac{Fo}{Fi} \text{ So, } Fo = \frac{3}{4} \times Fi$$

$$\frac{2\text{nd}}{3\text{rd}} = \frac{6}{5} = \frac{S}{1} \text{ So, } S = \frac{6}{5} \times T$$

$$\frac{Fo}{S} = \frac{\frac{3}{4} \times Fi}{\frac{6}{5} \times T} = \frac{\left(\frac{3}{4}\right) \times 4}{\frac{6}{5} \times 7} = \frac{3}{\left(\frac{42}{5}\right)} = \frac{3 \cdot 5}{42} = \frac{5}{14}$$

57

56. Here is one way to look at this. Think of dividing a line into six segments (separated by five lines – we don't need more than that because the highest number is 300,000) each holding up to three digits. Therefore, you have three digits and five separations - a total of eight segments. That becomes eight choose three (8C3) or 8!/5!3! = 56.

Or $111000 \dots \left(\dfrac{6}{3}\right) 20$

$210000 \dots 2\left(\dfrac{6}{4}\right) 30$

$300000 \dots \left(\dfrac{6}{5}\right)$

58

The smallest number of races needed is seven.

Race five horses in five races to get the top 15 horses (top three in each of the five races). Let those 15 be represented by:

A1	B1	C1	D1	E1
A2	B2	C2	D2	E2
A3	B3	C3	D3	E3

Horse A1 is the winner of his group, with A2 and A3 following. That holds for each vertical group.

Now, race the winners of each of the first five races: A1, B1, C1, D1, and E1. For the sake of ease, let's say the order of finish was: A1, B1, C1, D1, E1. This makes A1 the fastest horse.

A1 is faster than B1 and B1 is faster than C1, etc. By deduction, this makes it possible to eliminate groups D and E in their entirety.

A1 is faster than B1 and B1 is faster than C1 – we can now drop C2 and C3. Now, seven horses remain.

A1 is faster than B1. B1 is faster than B2 – drop B3 – six horses remain.

We already know A1 is the fastest – so just race the remaining five horses to get the Place and Show horse (the top two finishers of the final heat will be the overall second and third fastest horses, respectively).

59

The point on the hoop will travel exactly twice the diameter of the larger ring or 112 inches/pi. For a look at this actually occurring, take a look at example a/b = 2 on the following link: http://mathworld.wolfram.com/Hypocycloid.html.

60

There are 80 pitchers present who are not left-handed. The key is to see that 1/4 of the lefties are equal to 1/3 of all pitchers. Since $1/4\ L = 1/3\ P$, let $3/4\ L = 3\ x$. That means $1/4\ L = x$ and $2/3\ P = 2x$.

61

$x = -26$

$$\frac{1}{3^3} \cdot 3^{59} \cdot \frac{1}{3^5} \cdot 3^{3x} = 3^{x-1}$$

$$\frac{3^{59}}{3^3} \cdot \frac{1}{3^5} \cdot 3^{3x} = 3^{51+3x} = 3^{x-1}$$

Since the bases are now the same, the exponents must be equal:

$$51 + 3x = x - 1$$
$$52 = -2x$$
$$x = -26$$

62

There will be 233 pairs at the end of a year. One pair existed until February 28th. Then there were two pairs. The following month there were three pairs, then 5, then 8, etc.

1, 2, 3, 5, 8, 13, 21, 34, 55, 89, 144, 233.

Date Number of Pairs

Jan 31 – 1
Feb 28 – 2
Mar 31 – 3
Apr 30 – 5
May 31 – 8
June 30 – 13
July 31 – 21
Aug 31 – 34
Sept 30 – 55
Oct 31 – 89
Nov 30 – 144
Dec 31 – 233

63

C) 5.

1) $4^{19} - 2^{34}$
2) $4^{19} - 4^{17}$
3) Factoring $4^{17} \times (4^2 - 1)$
4) $4^{17} \times (5 \times 3)$
5) So 5 is the largest prime factor of $4^{19} - 2^{34}$.

64

The minimum number of cubes is 7.

65

D) 11

396 factors into $2^2 \times 3^2 \times 11$. So, $2^2 \times 3^2 \times 11^2 =$ 4356, which is 66 squared.

66

The ratio is 2 to 3. Let the bicycle's current age be $3x$ making the tires' age x when the bicycle was as old as the tires are now. To make them the same age we must add to the tire's age some number, y, and subtract from the bicycle's age the same number, y:

bike's age   tire's age
$$3x - y = x + y$$
$$2x = 2y$$
$$x = y$$

Since we've already established that $x = y$, we can substitute y for x in the bike's current age:

$3x = 3y$

The tires' current age is then $2y$, and the ratio of the tires' current age to the bicycle's current age is $2y/3y$, a ratio of 2 to 3.

67

Start by eliminating the width of the river. Draw a straight line for the rest of the distance. This can be accomplished by shifting City 1 downward by the width of the river and connecting it with a line to City 2 as if the river wasn't there. Now, put the river back, keep the line from City 2 to the edge of the river and put the bridge at the position, connecting its upper point to the original position of City 1. This is the shortest path.

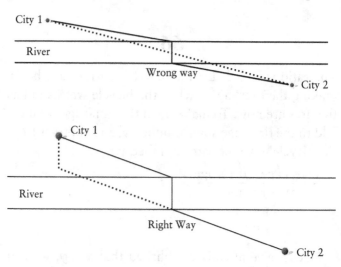

68

Frank is 28 and Sam is 21.

1) $x + y = 49$

2) $x = 2(y - (x - y))$

2a) $x = 2y - 2x + 2y$

2b) $3x = 4y$

$x + y = 49$

$3x - 4y = 0$

Multiply the first equation by 4

$$4x + 4y = 196$$
$$\underline{3x - 4y = 0}$$
$$7x = 196$$
$$x = 28$$
$$y = 21$$

69

11.57 units. Since A, B, and C are collinear, the slope of line AB will be the same as the slope of BC. Let S = Slope. Slope of AB is 2/7. Slope of BC is 2/7 = S/9. $7S = 18$; $S = 2.57$. Now, add 2.57 to 9 units to get the side length of Square 3... 11.57 units.

70

$\dfrac{112}{243}$.

The probability of five days where the machine is working is $\left(\dfrac{2}{3}\right)^5$ or $\dfrac{32}{243}$.

The probability of one day of the machine not working and four days where it is working is $\dfrac{1}{3} \times \left(\dfrac{2}{3}\right)^4$ and this can occur five different ways. That is $\dfrac{80}{243}$.

The probability of the new machine working on at least four days out of five is the total of the two... $\dfrac{112}{243}$.

71

The smallest value of $x \times y$ is $\frac{1}{2}$.

$$x + \frac{1}{y} = 12; \quad y + \frac{1}{x} = \frac{3}{8}$$

$$x = 12 - \frac{1}{y}; \quad y + \left(\frac{1}{12 - \frac{1}{y}} \right) = \frac{3}{8}$$

$$y \left(12 - \frac{1}{y} \right) + 1 = \frac{3}{8} \left(12 - \frac{1}{y} \right)$$

$$12y - 1 + 1 = \frac{36}{8} - \frac{3}{8y}; \quad 12y = \frac{36}{8} - \frac{3}{8y}$$

Multiply everything by $8y \rightarrow 96y^2 = 36y - 3$

$96y^2 - 36y + 3 = 0$

This factors into $3(8y - 1)$

$8y - 1 = 0$

$y = \frac{1}{8}$

$4y - 1 = 0$

$y = \frac{1}{4}$

$y = \frac{1}{4}$ and $y = \frac{1}{8}$

when   a) $y = \frac{1}{4}$, $x = 8$

　　　　b) $y = \frac{1}{8}$, $x = 4$

$\dfrac{1}{8} \times 4 = \dfrac{1}{2}$ is smaller than a) $\dfrac{1}{4} \times 8 = 2$

72

The mass of the contents when half the bolts are then removed is 54g.

mi = initial mass of contents

mf = final mass of contents

B = initial mass of bolts

N = initial mass of nuts

Solution: $mi = B + N$

But $B = 2N$

Therefore, $mi = 3N$

After removing half of the nuts

$$90 = 3N - 0.5N$$
$$90 = 2.5N$$

Half the bolts are then removed.

$$0.5B = N$$
$$mf = 2.5N - N = 1.5N$$

Since $2.5N = 90$, $1.5N = 54g$

$$\left(\dfrac{1.5}{2.5} = \dfrac{mf}{90} \right)$$

73

The daughter, Amira. If the best angler's twin and the worst angler are of the opposite sex, then the best angler and worst angler are of the same sex. The only possibility for this is the sister-daughter combination. Since the daughter and sister are the same age, the daughter must be a twin; the sister couldn't be or she, the daughter and the father would all be the same age! Therefore, the daughter, Amira, is the best angler and Amire, the sister, is the worst angler.

74

One way to think about this is to realize that it is Molly who is arriving at random times, not the trains. Imagine the trains from Miller Square to Murphy Center run every ten minutes and the same for Murphy Center to Miller Square... except each train bound for Murphy Center arrives one minute after the train bound to Miller Square. If Molly's random time falls between the time of a Murphy Center-bound train and a Miller Square-bound train (9 minutes), she will see the Miller Square train. If her arrival time falls between the time of Miller Square-bound train and a Murphy Center-bound train (1 minute) she will see a Murphy Center train.

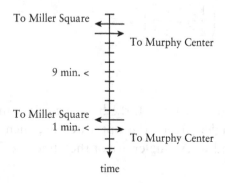

75

One rocket is three times as fast as the other.

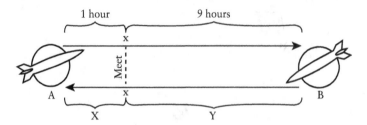

$X + Y$ = total distance

V_f = velocity of faster rocket

V_s = velocity of slower rocket

T_b = time before meeting

Y = velocity of the faster rocket multiplied by the time before they meet ($V_f \times T_b$)

X = velocity of the slower rocket multiplied by the time before they meet ($V_s \times T_b$)

Therefore, $\dfrac{X}{Y} = \dfrac{V_s}{V_f}$

Now after the rockets meet, Y is equal to the slower velocity multiplied by 9, and X is equal to the faster velocity multiplied by one.

Thus: $\dfrac{X}{Y} = \dfrac{V_f}{9V_x}$

We now have two different fractions that represent $\dfrac{X}{Y}$, and they are equal.

$$\frac{V_s}{V_f} = \frac{V_f}{9V_s}$$
$$V_f^2 = 9V_s^2$$
$$\sqrt{V_f^2} = \sqrt{9V_s^2}$$
$$V_f = 3V_s$$

76

b) 25 × 25 to the 500th power.

The first number is $5^{4000} \times 2^{2000}$, the second is $2^{2000} \times 5^{2000}$.

So, the correct multiplier is 5^{2000}, which indeed is $(25 \times 25)^{500} = (5^4)^{500} = 5^{2000}$.

77

60 miles per hour.

It takes Molly 21/12 or 1.75 hours to go the first 21 miles. For her to average 20 miles an hour for 42 miles, she needs to cover those 42 miles in 42/20 or 2.1 hours. That means she has to cover the last 21 miles in 2.1 – 1.75 or .35 hours. 2.1 ÷ 35 = 60. She will have to average 60 miles per hour on the second half to average 20 miles per hour for the entire trip.

78

They were charging approximately 12.5% more than their competition! Although it may appear that 1/45 oz. for $45 is an equitable midpoint between 1/30 oz. and 1/60 oz., it is not. 1/45 is 0.0222 while the mean of 1/30 and 1/60 is 0.02500. The average of the reciprocals of two numbers is different from the reciprocal of the average.

79

1. Pull entire train into BC segment.
2. Uncouple cars 3 and 4; leave them on BC.
3. Pull cars 1 and 2 over switch B and back them north on segment BA all the way over switch A.
4. Uncouple cars 1 and 2 and leave them north of switch A.
5. Bring engine down AC over switch C and back up on segment BC. Couple into cars 4 and 3.
6. Pull cars 4 and 3 over switch C, back them north on AC, and couple into cars 1 and 2.
7. Pull entire train down AB and leave west through switch B.

80

Robbie killed Fontani. Since each person could have only one false statement, and each maintained his innocence, that would be the killer's one falsehood. So, anyone who blamed someone else could not be guilty. If he were guilty, then claiming innocence would be a lie, as would be saying that someone else was guilty, making two false statements. Robbie is the only one who did not blame someone else, so he is the guilty one. He used someone else's gun ... and Benny and Doogie are not old pals.

81

18 square units. Take PT as the base of $\triangle$ PTR, then the height is 18 in. (RS). To find the base, notice that triangles PTQ and RST are similar. Ratio PQ : RS = Ratio PT : TS $\rightarrow$ 6 : 18 : : 2 : 6. That makes PT as 2 units. $A = 1/2bh$. $A = 2 \times 18 \times 12 = 18$ sq. units.

82

The missing face on the left is a 4 and the missing face on the right is 6. The touching faces on the right are the faces with a 2 on them. The touching faces on the left are faces that have a 5 on them.

83

a) 15/32. Here's one way to look at this:

4^{-1}	4^{-2}	4^{-3}		
$(1/4)$	$(1/16)$	$(1/64)$		
$\left(1 \cdot \dfrac{1}{4}\right) + \left(3 \cdot \dfrac{1}{16}\right) + \left(2 \cdot \dfrac{1}{64}\right)$			$= \dfrac{30}{64} =$	$\dfrac{15}{32}$ Base$_{10}$
1	3	2		

b) .14 Base$_{12}$

12^{-1}	12^{-2}	
1	4	
$\downarrow$	$\downarrow$	
$\dfrac{1}{12}$	$+ \dfrac{4}{144}$	$\rightarrow \dfrac{12}{144} + \dfrac{4}{144} = \dfrac{16}{144} = \dfrac{8}{72} = \dfrac{1}{9}$

c) 3.75

11.11 in Base$_2$ looks like this:

2^1	2^0	2^{-1}	2^{-1}	
1	1	1	1	
2	1	$\dfrac{1}{2}$	$\dfrac{1}{4}$	$\rightarrow 3.75$

d) $5797.531 \text{ Base}_{1/10}$

$$1 \cdot 10^3 + 3 \cdot 10^2 + 5 \cdot 10^1 + 7 \cdot 10^0 + 9 \cdot 10^{-1} + 7 \cdot 10^{-2} + 5 \cdot 10^{-3}$$

converted to $\text{Base}_{1/10}$ is:

$$5\left(\frac{1}{10}\right)^3 + 7\left(\frac{1}{10}\right)^2 + 9\left(\frac{1}{10}\right)^1 + 7\left(\frac{1}{10}\right)^0 + 5\left(\frac{1}{10}\right)^{-1} +$$

$$3\left(\frac{1}{10}\right)^{-2} + 1\left(\frac{1}{10}\right)^{-3}$$

84

There are nine ways (including the example):

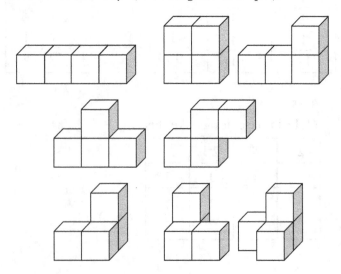

85

We found the twelve ways below. Can you find additional solutions?

$123 + 45 - 67 + 8 - 9 = 100$
$123 + 4 - 5 + 67 - 89 = 100$
$123 - 45 - 67 + 89 = 100$
$123 - 4 - 5 - 6 - 7 + 8 - 9 = 100$
$12 + 3 + 4 + 5 - 6 - 7 + 89 = 100$
$12 + 3 - 4 + 5 + 67 + 8 + 9 = 100$
$12 - 3 - 4 + 5 - 6 + 7 + 89 = 100$
$1 + 23 - 4 + 56 + 7 + 8 + 9 = 100$
$1 + 23 - 4 + 5 + 6 + 78 - 9 = 100$
$1 + 2 + 34 - 5 + 67 - 8 + 9 = 100$
$1 + 2 + 3 - 4 + 5 + 6 + 78 + 9 = 100$
$-1 + 2 - 3 + 4 + 5 + 6 + 78 + 9 = 100$

86

It will take 13 more minutes

The volume of a sphere is $V = 4/3 \ \pi r^3$. If it takes 30 seconds to blow up the balloon to 1 foot in diameter, then its volume is $4/3 \ \pi (1/2)^3 = 1/6 \ \pi$ (the radius 1/2 foot). For 3 feet in diameter, the volume will be $4/3 \pi (1.5)^3 = 4/3 \cdot 27/8 = 4 \cdot 5 \pi$. That is 27 times the volume than when the diameter is 1 foot. So, the total time it takes to fill the balloon is 27×30 seconds or $27 \times .5 \min = 13.5$ minutes. Since we have already taken 30 seconds to get the balloon to 1 foot, it will take $13.5 - .5$ or 13 more minutes.

87

Figure D. Its orientation is reversed from the other four.

88

143. Their system is based on something other than 10. Let's say it is based on N.

$2N + 6 = 26$, which is our 24 ($4 \cdot 6$). So, $2N + 6 = 24$; $2N = 18$, $N = 9$. Let's check our 30 ($6 \cdot 5$), their 33. $3N + 3 = 30$. Again, $N = 9$. Their number system is based on 9, and $4 \cdot 5 \cdot 6$, our 120, is their 143.

9^2	9^1	9^0
1	4	3

$= 81 + 36 + 3 = 120$

89

It will look like D.

90

The sixth view would look like:

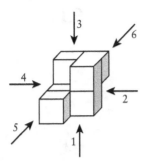

91

"B" would be placed 6th and "H" would be placed 4th.

D = 1	A = 5	I = 9
E = 2	B = 6	
C = 3	F = 7	
H= 4	G = 8	

92

.661 and .397, respectively.

In the first case, six out of seven people can be born on any other day of the week. So for seven people that, would be $(6/7^7)$ which is .3399. The certainty that someone is born on a Friday is 1. Therefore 1 – .3399 = .6601·

In the second case, the probability is $(6/7)^6$ or .3966. If only the first person were born on a Friday, the probability is $6^6/7^7$. But the probability is the same for each of the seven people. So, $7 \times (6^6/7^6)$ or $6^6/7^6$, which is .3966 (.397).

93

A is 3
B is 1
C is 4
D is 2.

94

123.

95

16 yards. The mistake the rabbits of Lepusville were making in their calculations was that they were assuming the fox and the rabbit were always in line on the same side of the hole. So, if the rabbit travelled farther than 48 yards, it had no chance for survival.

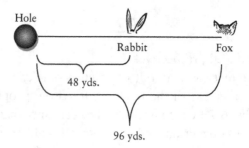

The diagram showed that the rabbits should be safe at any distance up to 48 yards from their hole as long as they were half as far away as Mr. Fox (which would put him at 96 yards). And yet, they were still being caught.

What Dr. Harvey knew was that one had to consider all possibilities of the different locations of both the fox and each rabbit. He then drew a second chart that brought everything into perspective for the citizens. Now, they knew they could never be caught by the wise 'ol fox. Here is the chart that saved their lives:

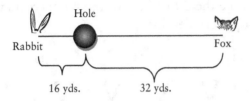

96

26 times.

A	1	F
B	1	C
C	2	BD
D	2	CI
E	1	L
F	1	A
G	3	HKJ
H	2	GK
I	3	DKL
J	2	GK
K	5	GHIJL
L	3	EIK

97

a) 1. This is the fraction 1/7 expressed in decimal.

b) 93. This is really pi - 2 digits at a time... 3.141592653589793.

c) –4. Take the differences of the numbers and you'll see a pattern develop.

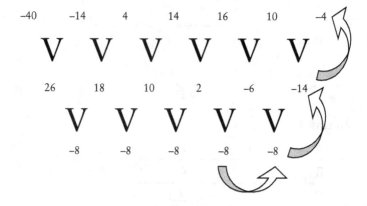

d) 3. Working from the outside numbers and moving toward the middle, each pair adds up to 19.

e) 144. This is the sequence 1, 3, 4, 8, 16, 32 and 64 in our number system of $Base_{10}$ converted to $Base_6$.

f) 14. This is a sequence where letters of the alphabet correspond directly with the counting numbers → A = 1, B = 2, C = 3, D = 4, and 0 = a space between words. The phrase spells PUZZLE FUN.

98

The weights from left to right:
40, 20, 10, 50, 30, 60, 70

Notice that the distance AC is 3 units as is the distance FG. That means the weight under AC must equal the weight under FG. We also have to keep in mind that for the entire system to balance, the weight times the distance under each segment must be equal on both sides of the entire mobile. The 2 weights under B must be in a 2 to 1 ratio because one arm is twice as long as the other. That means they are 10 and 20, or 20 and 40. They can't be larger because the largest single weight left (to be placed at G) is 70 pounds. So, the weights 10, 20, and 40, in some order, go in the three boxes under A and B. That means the 70-pound weight goes under G. Those 70 pounds are at a distance of 3 units from F; therefore, from A to C we have 70 pounds over 3 units for 210 ft. lbs. Even though these two segments, A to C and F to G, are equal, they are not in balance with the entire system. The weights of 50, 30, and 60 must be placed in the correct positions for everything to balance and one of the weights will not enter into any calculation of balance because it hangs directly in the middle, so neither adds or subtracts from either side. Under AC, we have either 30 + 60 or 90 ft. lbs. or 120 + 30 or 150 ft. lbs., depending upon how we place the 10, 20, and 40-pound weights. We need AI to balance with IG. Of the weights 10, 20, and 40, it can be seen that A is going to need the most weight. So place 40 at A, 20 at B_1, and 10 at B_2. That leaves 30, 50, and 60. It can be seen quickly through trial and error that the three weights will be 50, 30, and 60 from left to right. Check the numbers: From A to I, we have 280 + 150 + 50 or 480 ft. lbs. From I to G, we have 60 + 420 or 480 ft. lbs.

99

$2^{2^{22}}$ This is $2.065 \times 10^{1262611}$

100

```
    F          F
      F    F
        F
      F    F
    F          F
```

H is the 8th letter of the alphabet, L is the 12th. That is in the ratio of 2 to 3, as is B is to C. Likewise, P is the 16th letter of the alphabet, X is the 24th ... a 2 to 3 ratio. D is the 4th letter and F is the 6th letter to complete the 2 to 3 ratio across the board.

101

Place your X in the upper middle box. O can't win regardless where the next O goes.

AFTERWORD

"My object in living is to unite
My avocation and my vocation
As my two eyes make one in sight."
                                    —Robert Frost

For many years I have been fortunate enough to make Frost's quote a reality. I can't imagine earning a living doing anything else. I tried and was average (if that) in several other jobs. I didn't like any of them. But when I wrote my first puzzle column for the grand sum of $7.00 it felt like I had won the lottery. I was determined to make it work for the long haul. And I was just stupid enough not to realize the odds of making this happen are astronomically against it. I was also lucky enough to meet some special people along the way - people who went out of their way to help a fledgling writer. All these years later, they are still there offering suggestions and insights I would have never considered.

What a curious sounding thing to say when asked what I do for a living: "I am a puzzle writer." The reactions I get are hysterical. They range from complete dismissal to total misunderstanding to a body drain like an alien had sucked the brains and life out of the former human standing in front of me. Every time I am asked, I brace myself for their reaction, hoping I don't have to try to explain too much. How many careers offer this type of entertainment?

I would like to tell you that writing puzzles is hard work but that wouldn't be entirely true. While there are many things to juggle when attempting to create a new puzzle for someone's enjoyment, that becomes part of the game ... and is no different than the thrill of a batter trying to hit a pitcher's best. But I have several advantages over the hitter. I can break new ground as often as I please. I laugh all day long while writing. I get to meet interesting, bright people from the world over. I can write in my underwear. (Ask Derek Jeter to try that!) And no one will ever throw at my head.

What makes a good puzzle? It has to be both challenging and fun. I will let you in on a secret: it has to please me. In the end, I write puzzles that make me happy. I once read that great authors and musicians write to please themselves. While that seems overly self-absorbed, it seems to produce the most pleasing results. You have to be self-absorbed to think you can create something that will make other people happy. But there is a fine line to be drawn here. The truly great creators in all fields know when not to take themselves too seriously. There is always tomorrow for these people and that means they have to try to improve on today. You're only as good as your last at bat ... so get over yourself. Besides, puzzles are meant to be fun. This is not serious business for anyone other than the puzzle writer, but even we tend to keep it lighthearted most of the time.

People ask how I come up with the puzzles. I don't really know. I try to see things from different perspectives and put things together that normally don't fit. Sometimes, I try to be as silly as I can be, but not so silly that deductive reasoning is thrown away. I try to muster all the mental flexibility and creativity I can ... but contain it in a jar. Being a puzzle writer is not a career that one chooses. It chooses you.

I started all this at age 11, along with a friend, Tom Pyper, who was better at solving and creating than I. That's where the challenge part became so important. The fun was always there. I kept those early puzzles and they eventually became a book. The book did well and my puzzle writing career took off. I have been lucky enough to keep it alive for many years and I have never had a bad day. It just keeps getting better. And Tom? He went on to have a spectacular career as a brilliant trial attorney. I send him puzzles from time to time and he knocks them out in nothing flat.

Some days just slide by easily and the puzzles spring forth like they came from Zeus's head. Other days involve a little more sweat equity, but the days are never a grind. People ask me if I am running out of ideas. I haven't even scratched the surface. I can't wait to wake up each day and walk 30 feet to my office to write. Warren Buffett is quoted as saying "Find your passion and take the job you would take if you were independently wealthy." I have done that and never looked back. And I can't wait to see what tomorrow will bring.

Terry Stickels
www.terrystickels.com